AIR POLLUTION: ITS EFFECT ON THE URBAN MAN AND HIS ADAPTIVE STRATEGIES

by

KAIMAN LEE, AIA

EDRC OCCASIONAL PAPER

ENVIRONMENTAL DESIGN & RESEARCH CENTER
940 Park Square Building, Boston, Mass.

2-18-75

TABLE OF CONTENTS

CHAPTER I

INTRODUCTION

Man has long taken for granted his habitat and its apparently limitless bountifulness. Rather abruptly, and only recently, he discovered that not only were the bounties of his habitat limited but also that his imprint on that habitat could be seriously disruptive. Thus he came to realize that environment could not be taken for granted. Now he searches for processes that will ameliorate his environmental imprint. And he hopes it will not be too late.

Pollution of the atmosphere, the air we breathe in towns, residential and industrial areas, began many centuries ago and now in the 20th century has become one of the more important problems in many countries of the world. Since the population of the United States is about 70 percent urban, it is the cities where environmental problems converge and are noted most by planners and policy-makers.

Man is no longer seen as a weakling flinging up a panoply of technology to protect himself against the vicissitudes of some "hostile natural environment." Rather he is viewed as prime fouler of his own nest, the local urban scene, and also as affecting the world environment in significant ways.

My objective in this paper is to consider some concepts and actions that might contribute to the formulation of an adaptive strategy for air pollution.

CHAPTER II

HISTORICAL REVIEW

Since the beginning of the Industrial Revolution there has been an awareness of the potentially hazardous effects of air pollution. Concern about this problem has increased markedly over the years, and reached its height of expression in such activities as the National Conference on Air Pollution sponsored by the Surgeon General of the United States in December, 1962, and the establishment of an Expert Committee on Air Pollution by the World Health Organization in 1957.

Since 1930 a series of acute pollution episodes have occurred in scattered parts of the world; these episodes have demonstrated the danger and the lethality that may result when certain meteorologic phenomena occur in places where potentially high concentrations of air pollutants may form.

Belgium was covered by fog in December, 1930, during a period of anti-cyclonic weather. In the region of the Meuse Valley, 200 to 300 feet below the surrounding hills, the fog was especially dense. On the third day, many of the residents developed respiratory tract complaints. Persons of all ages were affected with throat irritation, hoarseness, cough, shortness of breath, and frequently a sense of chest constriction. Occasional nausea and even vomiting were reported. Cough seemed to be the predominant symptom, both productive and nonproductive. A number of those

affected developed increasing dyspnea, and some died. Those most severely involved were predominantly elderly people or those who had had previous chronic heart or lung disease. Retrospective study of this episode led investigators to indict sulfur dioxide.

In October, 1948, there was a large anti-cyclonic area in the northeastern United States, which was accompanied by fog and a prolonged weather inversion near Pittsburgh, Pennsylvania. Donora is about 30 miles south of Pittsburgh in a river valley surrounded by hills 400 feet high. During a 4-day period of fog, the inhabitants of Donora reported an increasing number of respiratory complaints. On the third day, 42.7 percent of the population (5,910 persons) complained of mild, moderate, or severe symptoms, which included smarting of the eyes, tearing, nasal discharge, sore throat, non-productive cough, nausea, headache, weakness, and occasional muscular aches and pains. A few reported vomiting and diarrhea. A total of 17 persons died.

A large area of the British Isles suffered anti-cyclonic weather from the 5th to the 9th of December, 1952. A heavy fog and an increase in air pollution occurred simultaneously, notably in the London area. Soon after the fog worsened, many people reported respiratory tract symptoms. Within the next few days, a large number of deaths occurred and an unusually high number of persons appeared at hospitals with complaints of productive cough and shortness of breath. Those who were most severely ill (many of whom subsequently died) were frequently those who had a history of

chronic cardiopulmonary disease.

Studies disclosed a correlation between sulfur dioxide and smoke levels and the number of sick.

A repeat of the 1952 episode appeared likely in December, 1962. This led to careful observation of pollutant concentrations and effects on health. Again, with the rising concentration of pollutants and the concomitant fog, large numbers of persons called for medical services. During the 1962 episode there were from 350 to 750 excess deaths--figures markedly lower than those in the outbreak 10 years earlier. At the same time, it seemed that the fog and weather conditions were equally severe, but that the chief difference lay in the quantity of pollutants. Specifically, although sulfur dioxide concentrations appeared similar to those observed in 1952, smoke was markedly reduced, perhaps to one-third of the 1952 level. Even though these data are not in themselves conclusive, they suggest that it is not a single pollutant, but a combination --this case, sulfur dioxide and smoke--that is of real significance.

During a prolonged weather inversion over the northeastern United States in November, 1953, fog and excessive air pollution occurred in New York City. From the incomplete data, it appears that there were an increase in hospital admissions and almost 250 excess deaths during that period. The vast amount of air pollutants and the great variety of questions related to atmospheric pollution have formed the object of studies by many different experts: physicians, chemists, botanists, engineers, factory

operators, smoke control officers, attorneys, public relations officers, and others.

CHAPTER III

THE ATMOSPHERE

Man lives at the bottom of the ocean of air that is the earth's atmosphere. He is so peculiarly adapted to it that he cannot exist at even moderate vertical distances from sea level without artificial help. He cannot exist indefinitely at altitudes exceeding about 5,000 meters without supplementary oxygen. Even moving downward in mine shafts presents problems to him as temperatures increase to dangerous levels. Thus it is the conditions that exist at and near the boundary between earth and atmosphere that provide the tolerable milieu for man's existence.

The composition of a dry atmosphere at sea level does not change much from day to day or from place to place over the earth's surface. Nitrogen (N_2) accounts for 78.084 percent; oxygen (O_2) accounts for 20.946 percent and Argon (A) accounts for 0.935 percent, while Carbon Dioxide accounts for almost all of the remaining 0.03 percent. The seven other gases encountered at concentrations of 10^{-3} and 10^{-18} parts.

Water vapor in the atmosphere varies from 0.02 to 4.0 percent by weight. Other substances in the atmosphere come from volcanic emanations: the burning of fossil fuel, smelting processes, decomposition of organic substances, various chemical interactions, salts derived from bursting bubbles at the ocean's surface, dust from wind and meteorites, radioactive fallout, and natural sources

such as pollen and other plant spores.

All of the additional air-borne materials--liquids, solids, and gases--excluding water in its several states, which may have adverse effects on humans and other kinds of life or which interfere with the enjoyment of life and property, are referred to as pollutants or contaminants of the atmosphere, thus the cause of air pollution.

Air pollution is not caused by a single particulate or gas in the atmosphere, but several. Carbon monoxide (CO), sulphur dioxide (SO_2), Nitrogen dioxide (NO_2), Ozone, ash, and asbestos all contribute to air pollution.

It has been demonstrated that man can live in an atmosphere of differing compositions. In the Mercury spacecraft, the atmosphere was 100 percent oxygen. Other atmospheres such as 50 percent oxygen and 50 percent nitrogen, oxygen plus helium, neon or argon are being tested. These experiments evidence that man adapts to his atmospheric environment.

CHAPTER IV

SOURCES OF AIR POLLUTANTS

Atmospheric contaminants are natural or are the product of man's activities. Among the natural are those arising from the soil, the sea, or outer space. Air motions and winds cause changes in the earth's surface, since they carry mineral dust from deserts, dried up lake beds, weathered rock, and eroded soil, transporting it often over great distances. The wind also takes up salts from foaming waves and the splashing surfaces of water reservoirs. Air streams can lift such salts to high altitudes and transport them hundreds of kilometers into the interior.

The products of volcanic eruptions are thrown up into the atmosphere. They contain such gases as HCl, HF, and CO_2 and ash with considerable quantities of water-soluble substances.

Electric discharges in the atmosphere cause the formation of nitrogen oxides and ozone. The metabolic by-products of bacteria, spores, and pollen also find their way into the air, as do the gases CH_4 and H_2S, formed in the decomposition of organic substances.

Many other substances are introduced into the atmosphere by man. These anthropogenic substances are products and waste materials from human economic activities. Most important are the industrial contaminants, primarily those generated by the burning of coal or other fuels--sulfur dioxide, nitrogen compounds, and smoke ash. Also significant are pollutants that form in chemical

production and radioactive contaminations. Human activities can also increase the natural contaminants of the atmosphere. One special type of localized air pollution is due to tobacco smoking.

In an urban environment the most widely distributed and important of all inorganic gaseous pollutants is sulfur dioxide resulting from the combustion of bituminous coal. Depending upon the type of coal used, the combustion results in the production of from 0.4 to 5 percent sulfur, which is present in coal as organic sulfur, sulfate, or sulfide. In combustion, the sulfur of sulfates become ash, whereas organic sulfur and that from sulfides, "combustible sulfur", oxidize to SO_2. The sulfur dioxide concentration in the atmosphere depends mainly upon the number of industrial plants in any area that burn significant amounts of coal.

The rapid development of every type of industry, particularly of chemical industries, the growth of towns, leading to the establishment of huge conurbations with an endless flow of motor traffic, and the chaotic way in which these urban areas are developed, often without provision of special zones for location of industry, etc., all contribute to the continuing pollution of ambient air. For instance, in Los Angeles and Washington, D. C., more than 50 percent of the pollutants in the air in 1963 was ascribed to vehicular emissions.

The periodic occurrences of smog resulting from photochemical reaction between hydrocarbons and oxides of nitrogen discharged in the exhaust gases from motor vehicles which causes irritation of

the mucous membrances of the eyes, nose, and throat should also be recalled. Similar occurrences of smog have been observed in San Francisco, Washington, D. C., and New York City. The growing use of motor transport throughout the world leads to increasing hazards from vehicle exhaust, and phenomena similar to photochemical smog have been reported in other counties such as Italy, Japan, and Australia. Pilot studies have shown that in the summer when there is intense solar radiation, oxidants produced by photochemical transformations of motor exhaust gases could be also found in the air in Moscow, Baku, and Batum in the Soviet Union.

The U. S. Public Health Service estimated that in the 5-year period ending in 1961, 150 million tons of pollutants were emptied into the American atmosphere each year. The potentially hazardous air pollutants to human health as they occur in the United States are listed below in order of importance: Carbon monoxide, Sulfur Dioxide, Nitrogen Oxide, Ozone, Sulfur Oxide, lead, asbestos, hydrocarbons, and photochemical oxidants. This report will deal in detail with the first two pollutants (CO and SO_2) listed.

CHAPTER V

GEOGRAPHIC DISTRIBUTION

The air pollution problem occurs when large numbers of people live and work together in small geographical areas in which the size of the available air space is inadequate for the safe dissemination and dilution of resultant pollutants. Although large cities with many chemical factories usually suffer most from air pollution, there are some surprising comparisons. According to the U. S. Public Health Service, in 1961, Phoenix, Arizona, with fewer inhabitants than New Haven, Connecticut, had more than twice the amount of suspended solids in its atmosphere; Canton, Ohio, with less than two percent of the population of New York City had a virtually identical (and heavy) pollution.

The concentration of carbon monoxide in large cities is highest in the morning and late afternoon--peak hours of automotive traffic. Investigations in Los Angeles showed that the greatest concentrations of oxidants occurred at noon. This is thought to be related to maximum solar radiation. Industrial contamination varies with factory usage within the week and according to the season. The seasonal variation in sulfur dioxide content shows a maximum in winter and a minimum in summer, clearly indicating a relation to fuel burning for house warming. Fog and wind may change the concentration of pollutants and a strong directional wind can alter the local distribution of industrial fumes.

Numbers and amounts of airborne pollutants are reaching such high concentrations in the atmosphere that significant shifts in global temperatures are expected by the year 2000.

CHAPTER VI

PROBLEMS OF THE ECOSYSTEM

All living organisms are linked to one another and to their environment by complex flows of energy and matter. This linkage is an open system which is called the ecosystem. Energy continually flows into the system from the sun. The energy is captured by the green plants in the process of photosynthesis. The energy stored in these primary producers is then distributed in a hierarchal manner to the herbivores, carnivores, and ultimately, the consumers of detritus. Because the energy of the system is finally dissipated as heat, it must be continually replenished from the sun. Materials, such as carbon, nitrogen, and water, flow through this system cyclically moving from environment, through the biomass, and back to environment. These hierarchal flows of energy and matter constitute the metabolism of the ecosystem. The basic ecological problem posed by air pollution is that the detritus of man's technology--heat, particulates, gases, and aerosols--are discharged into the atmosphere at a rate greatly exceeding the natural processes for recycling wastes. Thus, there is a net and continuing accumulation of technological detritus in the air.

In 1960, 23.3×10^6 tons of sulfur dioxide were released into the air. Most of the waste derives from the combustion of coal. By 1980 the figure for sulfur dioxide may reach 36.0×10^6 tons. In Los Angeles alone, the daily discharge of sulfur dioxide is 455 tons.

Nitrogen oxides are emitted at the rate of 835 tons per day. Organic vapor losses each day amount to 2,755 tons. Each day some 10,660 tons of carbon monoxide are discharged into the air.

These emissions derive primarily from the combustion of fossil fuels in industry and in automobiles. To capture the energy stored in these fuels, a remarkable volume of air is utilized.

It is estimated that each year 3,000 cubic miles of air are used to release the energy from the fossil fuels consumed in the United States. Automobiles alone account for 21 percent of this volume of air.

The atmospheric accumulation of detritus has initiated a degradation of the ecosystem in two distinctly different ways. First, the mounting discharge of wastes into the atmosphere has led to what has been euphemistically called "inadvertent" modification of weather and climate. Second, some of the wastes have proved toxic for various living organisms. For some time man did not perceive the ecological consequences of the growing bulk of atmospheric detritus. When, finally, some of the consequences were perceived, he labeled them "inadvertent." Of course, this was just rationalization for lack of ecological foresight. Now, the problems of air pollution have become so complex that man must make very difficult behavioral decisions, decisions upon which his continued survival may well depend. Now it is imperative that he understand the ecological implications of air pollution so that he may formulate an effective adaptive strategy.

A brief ecological analysis of four problems stemming from air pollution will serve to emphasize the complexity of the situation facing man.

A. Inadvertent Modification of Weather and Climate

Because of thermal pollution, urban areas are "heat islands." Large quantities of watse heat are dispersed into the atmosphere. It is not unusual to find the air over cities several degrees Celsius above that of the rural surroundings.

Carbon dioxide tends to trap outgoing infrared radiation. During the past three or four decades, there has been an increase of 15-25 percent in the concentration of carbon dioxide in the atmosphere. It has been suggested that this green-house effect of carbon dioxide has contributed to the rise of the mean global temperature of about 1.6 Celsius since 1900. Extrapolating from these correlations, predictions have been that by 2000 A.D. the mean global temperature may have risen 4 Celsius.

Concurrently, the turbidity of the atmosphere has shown substantial increases. The increments have taken place not only in the air over cities but, more meaningfully, in the air over places far removed from cities. It has been shown that lead from automobile exhaust serves as a potent freezing nucleus. Lead in the atmosphere may lead to increased cloudiness. In recent decades the number of condensation nuclei in the air has increased by several orders of magnitude.

All these man-made alterations argue for a growing disturbance in the energy budget for the ecosystem. What direction the disturbance will take is difficult to foresee. It is possible that the water budget may also be altered. Turbidity can act to stabilize the atmosphere. Increased stability means decreased convective activity and decreased rainfall. It may be that megalopolis will eventually produce its own drought.

In February 1973, Science Digest published an article by Dr. Cesare Emiliani, a noted American geologist in which he warned, "If the present climate balance is not maintained, we may soon be confronted with either a runaway glaciation or a runaway deglaciation." A runaway glaciation would incur an ice-age takeover. "This could be helped along by massive air pollution in the form of smog which would shade the earth from the heat of the sun and accelerate the cooling process." Runaway deglaciation would be the result of a man-caused global heat wave. This would not have to be spectacular, for scientists estimate that a rise in the earth's annual temperature of 3.5° C would be sufficient to melt the nine million cubic miles of ices covering Greenland and the Antarctic in just a few centuries. This would raise the world's sea level, thereby flooding every coastal city.

Man may be contributing to this process by pouring carbon dioxide into the atmosphere from smokestacks, chimneys, and automobile exhaust pipes. At this point, the world's climatologists are agreed on only two things: that we do not have the comfortable

distance of tens of thousands of years to prepare for the next ice age or flood, and how carefully we monitor our atmospheric pollution will have direct bearing on the arrival and nature of the weather crisis.

B. Atmospheric Dispersal of Pollutants

Meteorological processes, particularly wind and turbulent mixing, tend to disperse emissions for sources of pollutants. There is one important meteorological condition which, however, impedes dispersal; in fact, it traps air pollutants so that they may accumulate to very high levels. This situation is the inversion. The inversion has been the principal meteorological element in most air pollutions disasters and alerts. It has been shown that those geographic regions where conditions favor frequent development of inversions will be just the regions where the human population will congregate in 2000 A.D.

C. The Green Plant

Plants, in fact, are better sensors of air pollution than most instruments. This sensitivity has serious ecological implications, for the green plants play a number of vital roles in the metabolic flows of the ecosystem: The green plants capture solar energy and store it in forms useful to consumers. The plants participate in regulating atmospheric CO_2 and oxygen; in fact, green plants are primarily responsible for maintaining an oxidative environment for

all living organisms. In spite of these basic ecological functions, man's onslaught on the green plants has extended beyond the impact of air pollution. In assessing the role of green plants in the fitness of the ecosystem, the impacts of the urban sprawl and the rapid expansion of networks of highways must not be overlooked. These invasions are progressively reducing the land surface covered by green plants.

D. Human Health

It has been estimated that the amounts of pollen, natural dusts, smoke (as carbon), sulfur dioxide, and industrial dust produced in 1953 in the United States were 1, 30, 5, 19, and 7 million tons, respectively. Carbon monoxide emission alone amounted to approximately 60 million tons. All of these have important implications on human health.

As mentioned before, several classic examples of health effects of acute exposure to air pollutants have been recorded. During pea soup fogs in London in 1873, 1880, 1882, 1891, and 1892, an increase in the death rate was noted.

A study of overall mortality in New York City in 1962 and 1964 showed five recurring mortality peaks following the periods of intense atmospheric pollution and temperature inversion.

There are indications that atmospheric pollution may be a contributing environmental factor in human lung cancer. Epidemiological data indicates a constant rise in the frequency of lung cancer

in cities compared with the country. This cannot be explained entirely by the difference in prevalence of the smoking habit among inhabitants of cities as opposed to inhabitants of the country.

Medical evidence has shown that inhaled air pollutants can cause and aggravate such respiratory diseases as emphysema, pneumonia, and bronchitis, and can affect the brain and central nervous system, body defense mechanisms, and other vital organ systems. Air pollution also has a deleterious effect on man's overall work efficiency, intellectual and emotional functioning, and behavior in all age groups. There is even evidence that the healing process may be substantially impaired, prolonged, or actually prevented by air pollution.

CHAPTER VII

ADAPTATION

During the course of terrestrial and organic evolution, adaptation was achieved between organism and environment, an adaptation which may be viewed as fitness of the ecosystem. By virtue of this reciprocal fitness, continuing stability was achieved. This was before man became the dominant of the ecosystem; this has changed the universe: he has become its dominant. He extracts from the ecosystem energy and raw materials not only to sustain his metabolic requirements and to provide clothing and shelter for himself but also to support his technological establishment. He has learned to manipulate the ecosystem so as to enhance its productivity.

Man's manipulations of the ecosystem have come to pose a serious threat to the stability of this system. The view was emerged in recent years that, to assure his survival, man must manage, not manipulate, the ecosystem. In the context of this view, the ecosystem becomes a renewable natural resource. Its natural stability bespeaks its capacity for renewal. For example, the metabolic cycles assure a continuing supply of nutrient energy and raw materials, and photosynthesis maintains the concentrations of atmospheric oxygen and carbon dioxide. Management thus becomes the keystone of adaptive strategy. The objective of management must be to augment the productivity to fulfill man's specific needs and requirements. This objective must be pursued ecologically so

that the essential but delicate stability of the ecosystem is maintained.

A further important consideration in evaluating human risk is the question, "Risk to whom?" Some have argued that the risk should be evaluated in terms of the most sensitive, the least adaptable segment of the human population. The criteria by which this segment can be identified have not yet been established; and it is for this reason that studies of the range of adaptability to diverse environmental circumstances must be undertaken. The "air pollution episode" is one epidemiologic expression of the acute, overt, and direct effect, and it provides several ecologically useful insights. We learn of the role of meteriological processes, of toxic levels of air pollutants, of predisposing pathological conditions, of differential susceptibility, and of time-course for acute morbid events. Such facts as these are each important elements in adaptive strategy, for they provide inputs into the decision process of weighing costs and benefits.

A concept of health that is most useful for adaptive strategy is to view health as "a process of continuous adaptation to the myriad microbes, irritants, pressures and problems which daily challenge man." According to this view, health exhibits ontogeny, i.e., it matures, reaches a maximum, and declines during the life sequence of the individual. Furthermore, this view of health subsumes adaptive plasticity or the capacity of the individual to adjust to changing environmental circumstances. Certainly there is

ample evidence to support the ontogenic characteristics of health. The problematic characteristic is adaptive plasticity. This plasticity is largely past-oriented. We know that man can adapt to experiences through which he has come during his evolution. We are uncertain whether he will be able to adapt to the environmental changes that he is producing. Our uncertainty stems primarily from the fact that the rate of environmental change has been so rapidly accelerated. In the long course of time, genetic variability, which is so great in man, may assure survival through the action of selection pressures. But since the environment is being altered so rapidly, the process that most concerns us is the adaptability of men now peopling the earth.

It may be said that adaptive strategy is a system of cognition and behavior maximizing the well-being and minimizing the hazards to survival of human populations. The concepts fundamental to this strategy are habitat, biological population, and behavioral codes. Habitat and population are linked in the concept of ecosystem. Behavioral codes translate man's perception of himself in relation to ecosystem into a set of rules for resource management, for the ecosystem is a resource. The atmospheric environment, a part of the ecosystem, is equally a resource. Consequently, for an adaptive strategy for air pollution to be effective, it must treat air pollution ecologically. Although this viewpoint does not simplify the strategic decisions, it does assure that risks are evaluated and hazards are minimized.

The adaptive strategy is prospective, but the time for decision is upon us. The projections for continuing growth of the human population and its technological expectations suggest that current management practices will not keep pace, that atmospheric pollution will become more intense and more widespread. The urgency for instituting effective regulations was emphasized by President Johnson, now deceased, in his message to Congress on "Protecting Our National Heritage." In that message he called for an Air Quality Act which would establish regional regulatory commissions, promulgate uniform air quality standards, and expand investigative efforts. This message is an enlightened judgment, for it recognizes the fundamental dichotomy in an adaptive strategy, viz., the regulation of industrial growth (the cost) so that the wastes of technology will not irreparably destroy the fitness of the ecosystem (the risk).

A. Physiologic Adaptation

The most important adaptation in view of air pollution is in the respiratory function. Particles smaller than a few microns in diameter can penetrate into the parenchymatous areas of the lung while larger particles cannot; they may sediment onto the epithelium of the tracheal-bronchial tree or naval pharynx, or be expired. Penetration and residence time, therefore, are the major determinants of particulate deposition.

The respiratory pattern, i.e., the temporal relationship of tidal volumes and air flows, can greatly modify the foregoing

considerations. For example, slow deep respirations can be shown to favor particle penetration and deposition into the alveolar regions, while rapid shallow respirations favor upper respiratory tract and anatomical dead-space deposition.

Gases and vapors follow the principles which govern alveolar and dead-space ventilation more precisely than do particles. Their fate, like "soluble" particles, depends to a great extent on their chemical and physical properties. The extent to which they penetrate into the respiratory tract does not affect their over-all absorption particularly, but may determine the ease or rate at which they may diffuse into capillary blood.

B. Physical Adaptation

Adaptation to the environment is often purchased at the price of considerable morphological and functional changes.

In terms of morphological change, it is known that adaptation to irritant gases takes the form of a catarrhal condition of the mucosa of the upper respiratory tract which causes the sense of irritation to disappear. This does not mean that the irritant gases do not continue to have a harmful effect on the deeper parts of the respiratory tract. Adaptation as a result of chronic exposure to toxic substances is of an obvious pathological nature. It must be borne in mind that, first of all, pollutants have a biological effect and that the determination of harmless concentrations of pollutants represents a biological problem of the interrelationship

between organisms and environment; it represents the determination of the physiological boundaries of adaptation of the organism.

From a biological point of view functional activity is a result of many centuries of activity of living organisms under particular conditions of existence on the earth, under conditions of adaptation to the environment and the hereditary transmission of acquired morphological and physiological properties. In the course of phylogenesis, the organism has established specific and nonspecific adaptive mechanisms of many different kinds. The specific mechanisms are mainly designed to deal with environmental conditions which change rapidly, such as those of heat and cold and light and darkness. They comprise the fine adaptive reactions of the physical and chemical heat regulation system, the adaptive mechanisms of the eye, etc. Among the mechanisms triggering off these reactions are the body's exteroceptors and interoceptors, which constantly stand on guard, reacting sensitively to changing environmental conditions and transmitting signals to various parts of the central nervous system, which controls the adaptive reactions.

However, the body may be exposed to the harmful effect of chemical irritants to an extent which, because it exceeds the capacities of the physiological protective mechanisms--adaptation, requires the mobilization of defensive reserves whose role it is to compensate for the disturbed functions. According to an author's thinking individual organisms, including human beings, survive in polluted air because they process physiological reserves. Any policy of

exhausting these reserves, i.e., of accepting pollution so long as no damage to ourselves and our environment is apparent, is a policy of brinkmanship, involving more risk than most of us would be willing to take. Consequently, determination of the boundary between physiological adaptation, which does not involve stress, and the possibility of pathological reactions occurring, which require compensation of the disturbed functions in order to maintain normal interrelationships between the organism and the environment, is a most important task. This task consists in defining the biological effects and hygienic significance of atmospheric pollutants, and the aid of scientists in many branches--hygienists, physicists, epidemiologists, chemists, toxicologists, etc., must be enlisted with a view of solving the problems involved. Why is it impossible to accept the statement that the atmosphere should contain concentrations of pollutants at and above which there is likely to be irritation of the sensory organs? Is such irritation dangerous? The answer to this question can only be in the affirmative.

C. Carbon Monoxide

Carbon monoxide is a universal pollutant. It is estimated that gasoline engine exhaust is the source of about 75 percent of the carbon monoxide in the urban atmosphere. Peak rush hour traffic brings carbon monoxide concentrations into the range of 50 to 250 parts per million in large metropolitan areas.

In Los Angeles, California, until 1963, it had been increasing

at the rate of about one part per million each year.

Although there is no conclusive evidence that serious effects result from the concentration normally found in the air above an urban setting, this rate definitely indicates disastrous effects of carbon monoxide to health in the foreseeable future. Heating systems are a major source of carbon monoxide poisoning in the home.

The possibilities for adaptation of man to carbon monoxide are biochemical, physiologic, and psychologic. Biochemical adaptation processes might include increased detoxification of CO by endogenous metabolism and changes in anaerobic versus aerobic metabolism rates. Physiologic adaptation processes include increased hemoglobin concentration and cardiac output and possibly increased plasma volume and alveolar ventilation. Psychologic adaptation could occur by learning new material in the presence of CO, relearning old material, or revising a task to adapt to altered temporal perception. The cost of biochemical or physiologic adaptation might be reflected in altered life span, incidence of illness, and encroachment on maximal work capacity reserves or biochemical or endocrine system reserves. The cost of psychologic adaptation might be accidents.

The probable effects of chronic exposure to low levels of carbon monoxide may be on human health, behavior, and performance. The importance of CO in the ambient ari lies principally in its ability to combine with hemoglobin (HB). The portion of HB present in the form of carboxyhemoglobin (COHB) cannot combine with oxygen

or carry oxygen from the lungs to body tissues. Mental performance appears to be impaired at 2 percent COHB above background level (0.4 percent)(Committee on Effects of Atmospheric Contaminants on Human Health and Welfare, 1969). Although normal persons can compensate for some increase in inspired carbon monoxide (through increased HB concentration in the blood, increased cardiac output, or increased capillary blood volume), those with tissue hypoxia or less than normal oxygen in alveolar air might be more susceptible to the effects of low levels. There is evidence of a slight increase in mortality among patients with myocardial infarction following sustained exposure to more than 10 ppm carbon monoxide.

Some investigations implicate CO or an agent capable of imparting complex psychophysiological activities and associate it with deterioration of various sensory, perceptual, and cognitive functions in man, i.e., disturbed brain-cell oxygen metabolism; impairment of functions occur earliest in the higher centers of the central nervous system or in those areas that control cognitive and psychomotor abilities.

A single human subject was repeatedly exposed to low concentrations of carbon monoxide in an airtight chamber (Esther M. Killick, 1948); each exposure was prolonged until the carboxyhemoglobin percentage attained a steady value. The degree of acclimatization was indicated by the diminution in severity of the symptoms during successive exposures to the same concentration of CO, and the discrepancy between the observed percentage of COHB at the end of

exposure and the percentage of COHB obtained when the subject's blood pressure was equilibrated with a mixture containing oxygen and carbon monoxide at the same partial pressure as in the alveolar air. Acclimatization was not accompanied by changes in the red-cell count, in the proportion of reticulocytes, or in the blood volume. The results of inhaling a measured volume of CO from a closed-breathing circuit confirmed the existence of acclimatization and demonstrated that no appreciable destruction of CO occurred in the body. The value of the constant for the partition of HB between O_2 and CO, as determined, remained unaltered as acclimatization developed, the data obtained provide indirect evidence in favor of the hypothesis that the lungs excrete or prevent diffusion of CO.

In five recent continuous exposure studies (Anthony A. Thomas, 1969) PSIA pure oxygen and mixed gas atmosphere with sub-human primates performing continuous and discrete avoidance tests on audio and visual cues during exposure to 55, 110, 220, and 440 mg of CO per cubic meter have indicated an unexpectedly high tolerance and perhaps the presence of some adaptive mechanisms. CO in concentrations of 125 ppm had no demonstrable effect on higher mental integrating and coordinated neuromuscular function in human volunteer subjects for three-hour single exposure periods. There were no demonstrable physiological changes or clinical abnormalities.

CO was evaluated from three different approaches and a relative paucity of detrimental effects were found under low-level exposure

exposure and the attenuation of [illegible]

[illegible] blood pressure [illegible] with [illegible] containing

oxygen and carbon [illegible] at the same partial pressure [illegible] the

affinity [illegible] was not accompanied by changes [illegible]

The [illegible] coefficient [illegible] of [illegible] or [illegible]

of [illegible] The results of [illegible] a measured volume of [illegible]

from a closed [illegible] the extent of [illegible]

[illegible] and demonstrated that [illegible] of [illegible]

occurred [illegible] The value of the constant for the partition

of [illegible] as determined [illegible]

as [illegible] developed, the data [illegible] provide [illegible]

evidence [illegible] of the hypothesis that [illegible]

[illegible] diffusion [illegible]

In [illegible] continuous exposure studies [illegible]

[illegible] pure oxygen and [illegible] in [illegible] human

[illegible] performing [illegible] and [illegible]

and [illegible]

[illegible] high [illegible] and

[illegible]

[illegible] effect on human mental

[illegible]

[illegible] There were

[illegible]

[illegible]

[illegible]

conditions, both acute and chronic (James Theodore, Robert D. O'Donnell, and Kennect C. Back, 1970).

Exposure to relatively high levels of CO (460 MGM CU M and 575 CU M) for 168 days had no apparent effect on animal survivability, growth rates, or clinical chemistry and failed to produce pathologic changes in the CNS. The most prominent effect due to CO was the marked erythrocytosis seen during chronic exposure. This undoubtedly represented an adaptation to chronic tissue hypoxia induced by CO. No performance decrements were found in humans during three-hour exposures to 50-250 ppm CO. It is concluded that if CO at these levels had an initial adverse effect, adaptive processes must take place early during exposure, and the compensatory changes override the initial CO effect.

Dr. John R. Goldsmith (1969) concluded in his Technical Report on Air Quality for Carbon Monoxide with the following:

1. No human health effects are expected to have been demonstrated for COHb levels of 0-1 percent since endogenous CO production makes this a physiological range.

2. The following effects on the central nervous system (CNS) occur at 1-5 percent COHb: (a) about an estimated 2 percent COHb, there is an impairment in time-interval discrimination; (b) at about 5 percent COHb, there is an impairment in the performance of certain psychomotor tests, and an impairment in visual discrimination.

3. The following cardiac and pulmonary function changes occur

at COHb levels of 5-10 percent: increased ventilation, cardiac output, systemic arterio-venous oxygen content difference, systemic oxygen extraction ratios, myocardial arterio-venous oxygen content difference, and coronary blood flow in patients without coronary heart disease. In patients with coronary heart disease, the utilization of lactate by the heart muscle ceases, and excess lactate appears in the venous outflow; the compensatory increase in coronary blood flow which normally follows myocardial anoxia is impaired.

Long-term experimental exposure of humans to CO may produce certain effects such as increased hemoglobin levels and hematocrits but the available data are inadequate to draw firm conclusions.

It would have been appropriate to have included in the Conclusions that because of the strong affinity of hemoglobin for carbon monoxide, even small amounts of this gas in the ambient air will cause an increase in the COHb level. This increase will be in the order of 1-1/2 percent, if the ambient concentration is ten parts per million (10 ppm) and in the order of 3 percent at 20 ppm, if the exposure is continuous over an extended period of time. The endogenous CO produces a level of about 0.4 percent COHb.

Dr. Goldsmith's report failed to indicate the effect of COHb level of up to 20 percent. Others found the symptoms under this condition being vague and nondescript, but phenomenons such as: mild frontal headache, general weakness, fatigue, lassitude, and drowsiness occurred.

Another study of acute CO poisoning indicated the following

symptoms: unconsciousness, severe sequelae, restlessness, confusion, disorientation, and amnesia. Recovery of patient seemed to be attained after five days, but after 24 hours, suffered major acute relapses.

All CO poisonings must be quickly treated with 100 percent oxygen, if possible, under hyperbaric conditions. Prolonged observation and hospitalization for not less than a month after the intoxication is necessary since apparent recovery may be followed by severe illness.

D. Sulfur Dioxide

Sulfur Dioxide appears to have a world-wide distribution. Some of this chemical enters the air from natural sources such as volcanoes, but almost all of the sulfur dioxide in the urban atmoshere stems from the sulfur present in fossilized fuels, which are used virtually universally to produce heat and energy.

It has been shown that normal persons who inhale sulfur dioxide for brief periods exhibit shallower and more rapid breathing. All measurements of pulmonary resistance showed an increase. This was greatest for pulmonary flow resistance on quiet breathing, intermediate for pulmonary resistance on panting, or for airway resistance, and smallest for total respiratory resistance. Pulmonary flow resistance showed no change in exposure to 1 or 2 ppm of sulfur dioxide, but increase 19 percent at levels of 4 to 5 ppm and 49 percent at levels of 8 to 19 ppm.

Sulfur dioxide has the biologic effect that cause cessation (through either temporary paralysis or complete death of cells) of the ciliary motion of the mucosal lining of the respiratory tract. This inhibition may allow bacteria or carcinogentic agents to enter, remain on the respiratory epithelium, and produce noxious results.

The presence of aerosol may sweep the sulfur dioxide molecules deep into the respiratory system which may cause respiratory illnesses.

N. R. Frank (1969) summarized a group of studies on the physiological response of the lungs to sulfur dioxide (SO_2). Although the reactions of the lungs to an irritant may be quite diverse, just one feature of this response was of concern here, namely the changes in airway calibre reflected in measurements of pulmonary flow resistance (PFR). The subjects were volunteers who appeared to be free of underlying pulmonary disease. The range of SO_2 in these experiments exceeds that which is generally encountered in urban air pollution, although industrial laborers may on occasion be exposed to higher levels. However, there appear to be susceptible persons who react to quite low concentrations of the gas. Moreover, it is possible that the low concentrations of SO_2 that are found in polluted air, may, in association with other irritant gases and particles that are present, be sufficient to impair the function of the lungs.

Fifteen workers employed by a small chemical shop in Tirgoviste, Rumania, manufacturing a laundry bleach were periodically examined and the physiopathological effects of exposure to high concentrations

of sulfur dioxide were recorded (G. Untea, G. Bercu, M. Paraschiv, and I. Stefanescu, 1970). The workers were exposed to variable concentrations of SO_2 (between 70.4 and 184 mg SO_2/CU M air) during their entire work shift over periods exceeding six years. All exposed workers manifested lacrimation, 85 percent of the workers had excessive salivation, 70 percent had nasal hypersecretion. Seventy percent felt a burning sensation on the exposed parts of the body as well as in the respiratory tract; 74 percent complained of a painful cough with excessive expectoration. Pulmonary ventilation of 75 percent of the workers was characterized by polypnea during the latter half of the work shift with recovery following the end of work. Lung X-rays of workers exposed for more than six years manifested a marked pulmonary transparency; 87 percent manifested bilateral accentuated hilar images. The same workers also manifested a swollen haryngeal mucous membrane and similar symptoms on the endonasal mucous membrane; 37 percent of the workers complained of tiredness, asthenia, vertigo, and sleeplessness. All of these workers developed a tolerance to SO_2 and were able to withstand SO_2 concentrations which in unaccustomed individuals would have led to fatal intoxications considering that a concentration of 120 mg/CU M air cannot be tolerated for more than three minutes without the appearance of acute pathological symptoms.

An analysis of deaths that occurred in Chicago each day during January 1966 (Virginia Brodine, 1971) indicated no significant rise in female deaths with increased air pollution but suggested an

increase in the death rate among males 55 years of age and over when sulfur dioxide was between 0.17 and 0.29 ppm. When the mortality and pollution data were related to race and socioeconomic data, a revealing picture emerged. The dividing line between increase and no increase in deaths continued to be between 0.17 and 0.29 ppm for whites and between 0.12 and 0.24 ppm for the high socioeconomic group. However, it was only when SO_2 was 0.34 or higher that a significant increase was shown in either the non-white or the low socioeconomic group. The results are interpreted as implying that high levels of pollution can precipitate illness or death in less susceptible, more resistant individuals. The most susceptible individuals, e.g., the poor whites and non-whites, are killed by a combination of illness and environmental factors including the usual levels of pollution in the areas where they live.

CHAPTER VIII

LEGAL AND TECHNICAL STRATEGIES OF CONTROL

For control to be effective, to ensure that it not only interrupts pollution but prevents further pollution, technical and legal strategies must be applied in parallel.

On what should legislation for pollution control be based? What requirements should be laid down in regard to the sources of pollution? To what degree should industrial effluents be purified to prevent pollution of the atmosphere? The answers to these and numerous other questions in regard to legislation must be based primarily on the effect on man of various concentrations of substances discharged into the atmosphere.

Should absolutely pure air be insisted upon? In other words, should the air have the same composition as under natural conditions as it may be found in an environment uncontaminated by man? Of course not, because there are no grounds for asserting that every deviation from the natural composition of the atmospheric air will have an unfavorable effect.

A. Legal Strategy

Man must now make a number of complex decisions for managing the ecosystem. The decisions must be made so that industry can have a basis for planning and action so that the fitness of the ecosystem will be preserved, indeed so that the survival of the human

species will be assured. These decisions must have behavioral codes. The success of these codes will depend upon two factors: first, man's perception of himself and second, the quality of the information transmitted to the politician so that informed decisions can be made.

Man is now educated and trained to have a very parochial view of himself. His loyalties are primarily to small groups such as family, club, church, and school. They can enlarge to state and nation. But now those loyalties must become international and focus on all mankind. Environmental pollution is a problem that confronts all men and all men must be participants in its solution. The realization of this fact is threatening, it requires concern for people beyond one's immediate kin, indeed for people not even born; it symbolizes legal regulation, loss of individual liberty, and invasion of privacy.

They are as important, but inadequately studied, aspects of human adaptability as are the physiological regulations that make for biological adaptation. One urgent aspect of strategic planning thus will be to enlarge man's perspective of his place in, and relation to, the ecosystem in order to assure its continued fitness. All too often, political decisions are based on inadequate or incomplete information. It is equally the responsibility of the scientist to speak out on these matters and the responsibility of the politician to seek out appropriate and adequate information. Ecologists have been negligent in this important regard. Ecology, as an organized

profession, is not in good condition to become
creased research. As a scientific discipline,
focal point. As a point of view, it is already
ordinating other sciences and this may be the mos
function in the long run.

Insofar as environment is concerned, we are interested in environmental configurations which will permit adaptability to function effectively. In the context of this discussion, the particular environmental configuration that is the focus of our attention is air quality. Here the baseline might simply be atmospheric air polluted no more than might be expected by natural processes--dust storms; sea spray; pollen from grasses, flower, and trees; and volcanoes. From the viewpoint of human health, how much of a departure from such a natural atmosphere can be accommodated by man's adaptive plasticity? The answer to this question is given in the decision man makes about "air quality standards." If the "air quality standards" are set so that some man-made pollution is permitted, there is a risk that a deterioration of health may accrue. Since very little is known about human adaptability for air pollution, the magnitude of the risk cannot now be assessed. Consequently, since enforcement of air quality standards will entail great cost, it is imperative that the risks be established so that the standards can be realistic. In other words, in our strategic thinking we must weigh the risk of man's adapting to a polluted atmosphere against the cost of regulating the atmosphere so that it fulfills man's needs and requirements.

Certainly, man's capacity to adapt to a wide range of circumstances deserves more explicit recognition in our environmental planning as an alternative to a policy of effecting adjustments of the environment aimed toward some assumed set of maximally satisfying conditions.

In the World Health Organization monograph, "Effects of Air Pollution on Human Health," Heimann pointed out that to maintain health it was not at all necessary that the air we breathe should contain no impurities whatsoever [18]. What is important is not so much the presence of impurities as their concentrations. We must know what levels of pollutants in the atmosphere are dangerous if inhaled. It is difficult to agree that such concentrations of pollutants in the atmosphere, which are toxic to man, but which are said to be practically impossible to eliminate at the present level of technical development should be accepted. Technology in the modern world is so highly developed that there are very few questions connected with the technology of atmospheric pollution control which cannot be solved from the theoretical and even the practical point of view.

In 1963, the World Health Organization's Expert Committee on Atmospheric Pollutants approved the basic conclusions reached by an inter-regional symposium on criteria for air quality and methods of measurement, sponsored by the World Health Organization. This committee suggested that guides to air quality may be presented for four categories, with the concentrations, exposure times, and corresponding effects for each, as follows [18]:

Level 1. Concentrations and exposure times at or below which, according to present knowledge, neither direct nor indirect effects have been observed.

Level 2. Concentrations and exposure times at and above which there is likely to be irritation of the sensory organs, harmful effects on vegetation, visibility reduction, or other adverse effects on the environment.

Level 3. Concentrations and exposure times at and above which there is likely to be impairment of vital physiological functions or changes which may lead to chronic diseases or shortening of life.

Level 4. Concentrations and exposure times at and above which there is likely to be acute illness or death in susceptible groups of the population.

What concentrations should be aimed at in order to prevent the harmful effect of atmospheric pollution on the health and well-being of man?

Level 1 is of primary practical interest for the control of atmospheric pollution, since any concentrations above that level would be inadvisable.

At the present time, as a result of air pollution control and abatement programs, the situation in a number of countries has begun to improve. Restrictions put on the sulfur content in fuel, the substitution of natural gas for mineral fuels, the electrification of industrial processes, the provision of effluent cleaning devices, and

other measure have helped in reducing the smoke problem both in theory and in practice.

B. Technical Strategy

We can protect the sanitation of drinking water by providing one or several barriers, such as protection of water sources and their storage; we can purify and disinfect water, but, owing to the mobility of gaseous substances, the only effective barrier that can be applied to air is the prevention of atmospheric pollution.

The diminution of pollutants is possible at industrial plants by the rational use of raw materials, the reduction of dust of powdery materials as in cement plants, adequate hermetic sealing in factories, and the use of dust removal apparatus, which may be mechanical, electrostatic, or sonic, depending upon the nature of the dust and the degree of dust removal indicated. The efficiency of dust-removal apparatus is related to the ratio between dust removal and dust supply; some equipment is designed to absorb dust of a certain size.

Of great importance is apparatus to neutralize gases before their escape from the stack. Condensation of the gases and their absorption by a suitable layer containing binding or neutralizing elements are accomplished by apparatus built into some equipment. It is incumbent upon management and the plant engineer to prevent, by fire box design and other methods, escape into the atmosphere of noxious gases, soot, and smoke.

Another major factor in reducing the volume of particulate contaminants in large cities is the preservation of cleanliness of the sidewalks and streets. City planning should include the rational distribution of green areas and factories to minimize concentration of pollutants, especially near residential areas.

A guide was established to aid in the selection of trees and shrubs adapted to the environmental condition with large amounts of sulphur dioxide (M. V. Bulgakov, 1969). An assortment of gas-resistant species for city plantings was developed, including poplar, birch, cedar, larch, maple, elm, Siberian pea shrub, elder, dogwood, sweetbrier, and others. Siberian pea shrubs from the local nursery and from another nursery were planted in a city park. The leaves on the plants from the local nursery remained normal in appearance, while those from the other nursery were severly damaged by SO_2. This confirms the fact that the trees and shrubs for gas-polluted areas should be exclusively local, raised in the same noxious medium where they are to continue their growth. If frost-resistance is also desired, the plants should be raised under rigorous, spartan conditions. The establishment of parks and plazas, and the great number of trees and shrubs along the streets have made the air of the city cleaner and resulted in a milder microclimate.

CHAPTER IX

FURTHER RESEARCH

An area of definite need in further research is in the measurement of air pollution concentration in a specified location. The concentration will depend on a vast number of variables. As the air pollutants emerge, they mix with the outside air and begin to disperse in accordance with existing meteorologic conditions--the velocity, direction, and frequency of variation of wind--which control the smoke plumes, the rising and settling of the particles, and the mode of their dispersion. The temperature also influences pollutant dispersion, The vertical fall of air temperature with increasing height--the so-called temperature gradient--is related to the season. In summer it is approximately 1°C. per 100 meters; in winter it reaches only thousandths of a degree, and sometimes the value is actually negative. Such a thermal inversion causes a deviation in the direction of flow of the pollutants. They may then concentrate in the ground air layer, which is also influenced by humidity. They may then become heavier with condensed water vapor and sink to the ground layer, where further concentration occurs.

The height of the smoke stack is very important. The higher the stack, the greater will be the utilization of wind velocity, the more complete will be the mixture, and the larger will be the cross section of the fume cone before it touches the ground, which means that the amount of soil contamination per unit of surface will

fall.

Atmospheric pollution changes in time and space. There have been many studies on the relationship between the distance of the source and the extent of dispersion. It is obvious that the greater the distance from the source, the more thorough will be the mixing of the pollutants and the air. Many observations on the angle of the smoke plume have confirmed this. Mathematical formulae for the determination of dispersion have failed, since there are so many natural factors that deform the smoke cone, such as equalization of smoke and wind velocity and obstructions on the ground and above it. Although much is known about local dispersion, very little is known about the movement of air pollutants over large countries or continents or the globe. Mathematical modeling and computer simulation may be the most feasible tools in solving this problem.

To understand fully the physiological effects of pollutants on the individual, one must have knowledge of their distribution in the environment, incorporation into the body, transmission or storage in the body, release from storage, access to body spaces such as cerebrospinal fluid, effects on nerve cells, manner of detoxification, and excretion. Research on some of the substances commonly thought of as pollutants has barely begun.

Several ways are open to obtain information on adaptive strategies. One approach would be to study populations representing different genetic backgrounds living in contrasting habitats and cultural circumstances. A second approach would be to identify in a

population the range of adaptability to diverse environmental circumstances. A third approach would be to investigate populations migrating between contrasting habitat and cultural circumstances. Each approach should be longitudinal, for adaptability can only be judged realistically by continuing observation. Each requires detailed and intensive study of the members of the selected populations. To fulfill the objectives of these studies, detailed measurements of the sort performed by clinical investigators must be carried out on representative populations. In a sense, this is human experimentation. It can be argued that since man is already at risk, but at risk of unknown magnitude, such measurements are justified, indeed imperative.

The social, physical, and psychological characteristics that explain the idiosyncratic responses require further study. A few such characteristics have been identified. Studies show, for example, that attitudes of people toward the source of environmental stimulation can affect how disturbed they are by this stimulation. Such a finding represents an important first step in the search for the individual characteristics that explain the variation in the outcome variable.

CHAPTER X

CONCLUSION

It has been shown that there are at least two potential weak links in the ecosystem: the green plant and the human being. The plant is the primary producer of nutrient energy for the hierarchy of dependent consumers including man. Thus from the viewpoint of an adaptive strategy, it must take into account the impact of air pollution on the green plant which is the weakest link in the metabolic flows of the ecosystem. Since human health is directly affected by air pollution, it is his adaptive strategies on which this paper has placed its emphasis.

One of the major causes of air pollution is population growth that asks for more and disposes more. In light of this dilema, it is believed that the most effective long term strategy is to control population growth; its goal is the balance of the ecosystem.

Legal strategies such as behavioral codes, air quality standards, and environmental impact statements for new construction projects are relatively convenient to set up but difficult to enforce. However, they are essential strategies because once legal limits of air pollution are set and enforced, the industry will have to find the suitable technological strategies to meet with these legal standards.

Technological strategies such as new types of fuel, new automobile engines, neutralization of gases before emission into the air,

and pollution traps are being developed to prevent air pollution. The high cost of doing these are vividly apparent. Man must be educated to treat the air pollution problem in an ecological manner so as to realize that such cost is justified. The temporal strategy to overcome acute air pollution can be by wearing masks and avoiding or escaping the source of pollution or ignoring it.

Since man's physiological adaptation is most effective at the outset of his air pollution encounter, and that the adaptive mechanism becomes less effective and will deteriorate through time, this physiological strategy cannot cope with the chronic air pollution problem.

Although we do not have the research nor the test results to prove the legal air quality standards, we must set these standards on the basis that we already know the existence of health defects of man caused by air pollution. Air quality standards can be set at arbitrary or best possible levels. They will serve as a means to test the results and to curb further air pollution, if not less, on our earth.

All of these strategies cannot operate independently, they must form a team to tackle this most imminently urgent problem before we realize that it is too late and we have had to sacrifice our values --including our lives.

BIBLIOGRAPHY

1. Amdur, M. O. The physiological response of guinea pigs to atmospheric pollutants. International J. of Air Pollut., London, Vol. 1, pp. 170-183.

2. Beard, Rodney R. Toxicological appraisal of carbon monoxide. J. Air Pollution Control Assn., September 1961, Vol. 19, No. 9, pp. 722-732.

3. Bonnerie, P. Atmospheric contaminants and human health. Interregional Symposium on Criteria for Air Quality and Methods of Measurement, Geneva, Switzerland, Aug. 6-12, 1963.

4. Brodine, Virginia. A special burden. Environment, March 1971, Vol. 13, No. 2, pp. 22-24 and 29-33.

5. Bulgakov, M. V. An experiment in creating protective plantings in the City of Krasnoural'sk. In: American Institute of Crop Ecology Survey of USSR Air Pollution literature. New York: Nuttonson (ed.), Vol. 2, Silver Spring, Md., American Institute of Crop Ecology, 1969, pp. 79-84.

6. Burns, Neal M., Randall M. Chambers, and Baldwin Hendler. Unusual environments and human behavior. London: Collier-Macmillan, Ltd., The Free Press of Glencoe, 1963.

7. Carson, Daniel. Environmental stress and the urban dweller (a study). Michigan Mental Health Research Bulletin, Winter, 1969, I, p. 15.

8. Carson, D. H. The interaction of man and his environment. Ann Arbor, Michigan: SER 2 School Environments Research--Environmental Evaluations, 1965, pp. 13-52.

9. Committee on Effects of Atmospheric Contaminants on Human Health and Welfare. National Academy of Sciences - National Research Council, Washington, D. C. Summary of tentative conclusions. in effects of chronic exposure to low levels of carbon monoxide on human health, behavior, and performance, 1969, pp. 3-6.

10. Dill, D. B., E. F. Adolph, and C. G. Wilber. _Handbook of Physiology_ (Section 4. Adaptation to the environment). Washington, D. C.: American Physiological Society, 1964.

11. DuBois, A. B. Adaptation to carbon monoxide. In: Report prepared by the Committee on Effects of Atmospheric Contaminants on Human Health and Welfare, National Academy of Sciences, Washington, D. C., entitled: Effect of chronic exposure to low levels of carbon monoxide on human health, behavior, and performance, 1969, pp. 38-39.

12. Dubos, Rene. Man Adapting. New Haven, Connecticut: Yale University Press, 1965.

13. Folk, Edgar G., Jr. Introduction to Environmental Physiology. Philadelphia: Lea and Gegiger, 1966.

14. Frank, N. R. Studies on effects of acute exposure to sulphur dioxide in human subjects. Proc. Royal Soc. Med. 57, (10, Part 2), October 1964, pp. 1029-1033.

15. Hall, Edward T. Human needs and inhuman cities. Ekistics, (Discussion), 27 March 1969, pp. 181-184.

16. Helvey, W. M. Physiological and psychological problems in space flight. Washington, D. C.: NASA Publication SP-205, 1967, Biotechnology.

17. Hine, C. H., and F. H. Meyers. The human subject and air pollution research. Air Pollution Medical Research Conference, Los Angeles, Calif., December 4, 1961.

18. Izmerov, Nikolai F. Establishment of air quality standards. Archive Environmental Health, June 1971, Vol. 22, No. 6, pp. 711-719.

19. Izmerov, Nikolai F. Some biological aspects of air pollution. World Health Organization Chronicle, Geneva, Switzerland, February 1971, Vol. 25, No. 2, pp. 51-57.

20. Killick, Esther M. The nature of the acclimatization occurring during repeated exposure of the human subject to atmospheres containing low concentrations of carbon monoxide. J. Physiol., London, 1948, Vol. 107, pp. 17-44.

21. Lee, Douglas H. K., and David Minard. Physiology, Environment, and Man. New York: Academic Press, 1970.

22. Light, Sidney (ed.). Medical Climatology. New Haven, Connecticut: Elizabeth Licht, 1964.

23. McDermott, M. Acute respiratory effects of the inhalation of coal dust particles. J. of Physiology, London, Vol. 162, p. 53.

24. National Institute of Mental Health. Pollution: Its Impact on Mental Health. Rockville, Maryland: National Clearinghouse for Mental Health Information, National Institute of Mental Health.

25. Otis, A. B. The physiology of carbon monoxide poisoning and evidence for acclimatization. Annual of New York Academy of Science, 1970, Vol. 174, pp. 242-245.

26. Pfrender, R. E. Chronic monoxide poisoning. Ind. Med. Surg., March, 1962, Vol. 31, pp. 99-103.

27. Platt, Robert B., and John F. Griffiths. Environmental Measurement and Interpretation. Feinhold College Textbook Department.

28. Proshansky, Harold M., William H. Ittelson, and Leanne G. Rivlin. Man in His Physical Setting. New York: Holt, Rinehart & Winston, 1970.

29. Ricci, C., F. Capellaro, and P. C. Gaido. Electrophoretic and Immuno-Electrophoretic examination in workers exposed to chronic carbon monoxide intoxication. RASS, Med. Ind., May-August 1964, Vol. 33, No. 3-4, pp. 414-416. (Translated from Italian)

30. Ryazanov. V. A. Sensory physiology a basis for air quality standards: the approach used in the Soviet Union. Archive Environmental Health, November 1962, Vol. 5, pp. 86-100.

31. Sabaroff, B. J. The Bio-psycho-sociological Effects of the Environment on Man. Providence, R. I.: Rhode Island School of Design, April 1966.

32. Sargent, Frederick, II. Adaptive strategy for air pollution. Bioscience, October 1967, Vol. 17, No. 10, pp. 691-697.

33. Sims, V. M., and R. E. Pattle. Effect of possible smog irritants on human subjects. J. of Am. Med. Assn., Vol. 165, pp. 1908-1913.

34. Smithsonian Institution Press. The Fitness of Man's Environment. Washington, D. C.: Smithsonian Institution Press, 1968.

35. Stern, Arthur Cecil. Air Pollution. New York: Academic Press, 3 Volumes.

36. Stokinger, Herbert E. The spectre of today's environmental pollution--U.S.A. brand: new perspectives from an old scout. Am. Ind. Hyg. Assn. J., May-June 1969, Vol. 30, No. 3, pp. 195-217.

37. Theodore, James, Robert D. O'Donnell, and Kenneth C. Back. Toxicological evaluation of carbon monoxide in humans and other mammalian species. In: Lectures in Aerospace Medicine, USAF School of Aerospace Medicine, Brooks AFB, Texas, Aerospace Medical Div., Seventh Series, February 9-12, 1970, pp. 194-232.

38. Thomas, Anthony A. Toxicology of carbon monoxide in artificial cabin atmospheres. AGARD Conf. Proc. (Paris, France), No. 61 - 3-1,3-2, 1969. Presented at the Aerospace Medical Panel Meeting, 26th, Florence, Italy, October 21-24, 1969.

39. Untea, G., G. Bercu, M. Paraschiv, and I. Stefanescu. Physiopathological effects and tolerance phenomenon connected with excessive atmospheric concentration of sulphur dioxide. Translated from French, Arch. Maladies, Profess. Med. Trav. Securite Sociale (Paris, France), September 1970, Vol. 31, No. 9, pp. 481-484.

40. World Meteorological Organization, Geneva, Switzerland, and International Society for Biometeorology. A survey of human biometeorology. TN-65, WMO-160, Tp. 78, 1964, p. 113.